The Race to Recycle

by David Meissner

illustrated by Stan Tusan

Max woke up early.
His eyes grew big.

Max ran to his sister's room.
"Tanya, get up. It's Monday!"

Tanya looked at the clock. "Hurry! The truck comes in five minutes," she said.

Max put the old newspapers in the bin. Tanya cleaned out the old bottles and jars.

Max found some juice left from the night before. "Cool!" he said as he drank the rest.

He put the can in the bin.

Tanya and Max ran to the street and looked around. The other bins on the street were still full.

"We made it!" Tanya said.
"The truck has not come yet."

"Tanya, what happens to these things?" Max asked.

"They are made into new things," Tanya said. "They get a new life."

"Look!" Max said. "Jenny's bins are not out yet! I bet she forgot to put them out."

Tanya and Max ran to Jenny's house. Max went up the tree by her window. "Hey, Jenny! It's Monday! You have to put your bins out!"

Jenny let her friends inside.
They put the bottles, jars, cans,
and newspapers into bins.
They were just in time
for the truck.

"Thanks for your help!" said the man. He put the bottles, jars, cans, and newspapers into the truck.

"Have a nice new life!"
Max called to the things.